Hands-On Projec

SPACE

Active Learning about the Solar System

by Carol Wawrychuk & Cherie McSweeney
illustrated by Philip Chalk

Contents

Lovingly dedicated to Bill Wawrychuk for his never-ending love—Carol

For a complete catalog, please write to the address below:
P.O. Box 1680, Palo Alto, CA 94302 U.S.A.
Call us at: 1-800-255-6049
E-mail us at: MMBooks@aol.com
Visit our Web site:
http://www.mondaymorningbooks.com
Monday Morning is a registered trademark of
Monday Morning Books, Inc.

ISBN 1-57612-041-4
Printed in the United States of America

987654321

Introduction

10-9-8-7-6-5-4-3-2-1-BLAST OFF! The *Space* theme unit is ready to bring the curiosities of the solar system to your children.

Through the use of a large appliance box space shuttle, youngsters become astronauts exploring the universe. A life-size astronaut, fashioned from old clothing and a pillow case, welcomes little astronomers to the dryer box planetarium.

Boys and girls have a hands-on introduction to the solar system as they create plaster of Paris moon craters, crayon-resist skies, and Jupiter sand jars. Children will be thrilled to make the Mars' rover which can travel on their "red planet" made from salt, flour, water, and paint. Letter and number recognition games complete the learning experience.

Whether children pretend to be astronauts walking on the moon or astronomers discovering distant planets, imaginations will soar with this unit on "Space."

Personal Observations:

As children eagerly helped assemble the space shuttle, they asked questions about the solar system. While stuffing newspapers in the clothing to make the life-size astronaut, children made suggestions for the astronaut's name. The space shuttle took center stage as children pushed it around the room to find safe places for a take-off or landing.

The glow from flashlights illuminated the inside of the planetarium where children gazed at the stars and planets. Some pretended to be astronomers giving a guided tour at an observatory. This activity was so popular that a cooking timer was used to ensure that each child had a turn.

Children told stories about space as they colored and cut out the paper doll astronauts and space shuttles. Some had their astronauts form a "team" to fly in one space shuttle.

All of the activities in the "Space" unit came together to make this an "out of this world" experience!

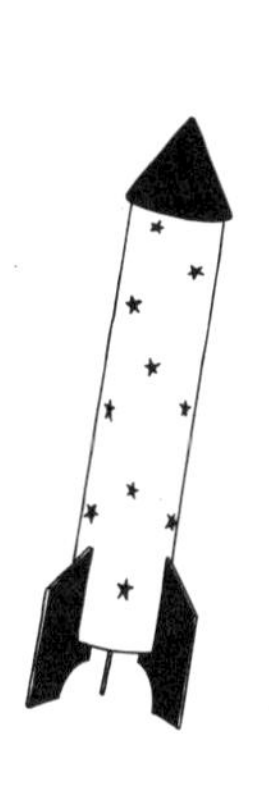
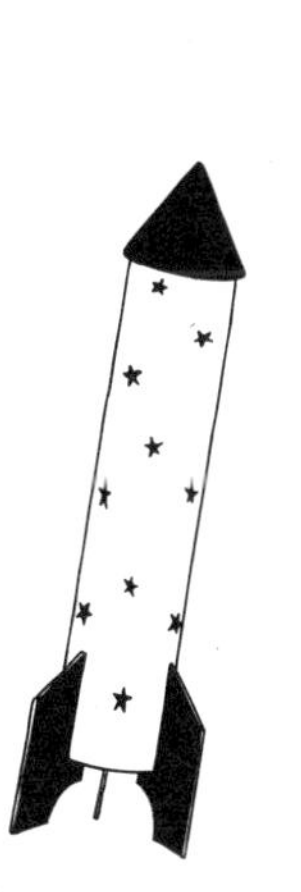
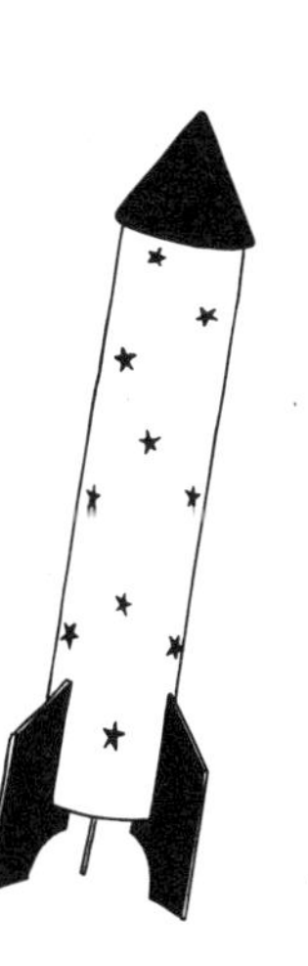
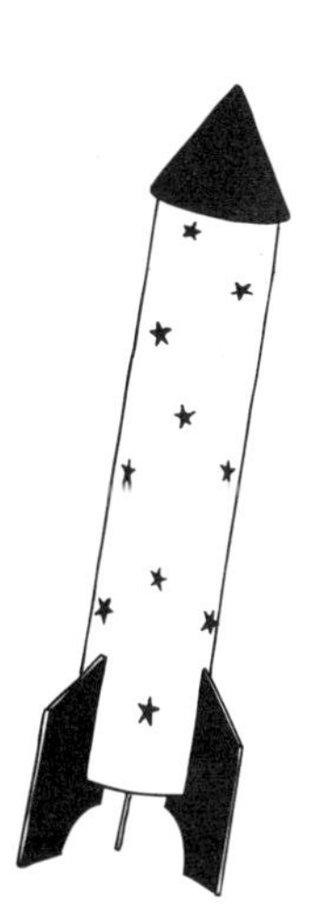

Space Shuttle

Materials:
Large box (the size of a four-drawer file cabinet), newspapers, red tissue paper, sturdy paper bowls, masking tape, glue, yarn, tempera paint (black and white), paintbrushes and rollers, shallow tins (for paint), papier-mâché paste (flour and water), sharp instrument for cutting (for adult use only)

Directions:
1. Seal the flaps on both ends of the box with masking tape, and lay the box on its side.
2. Cut wings in one side of the box according to the diagram below.

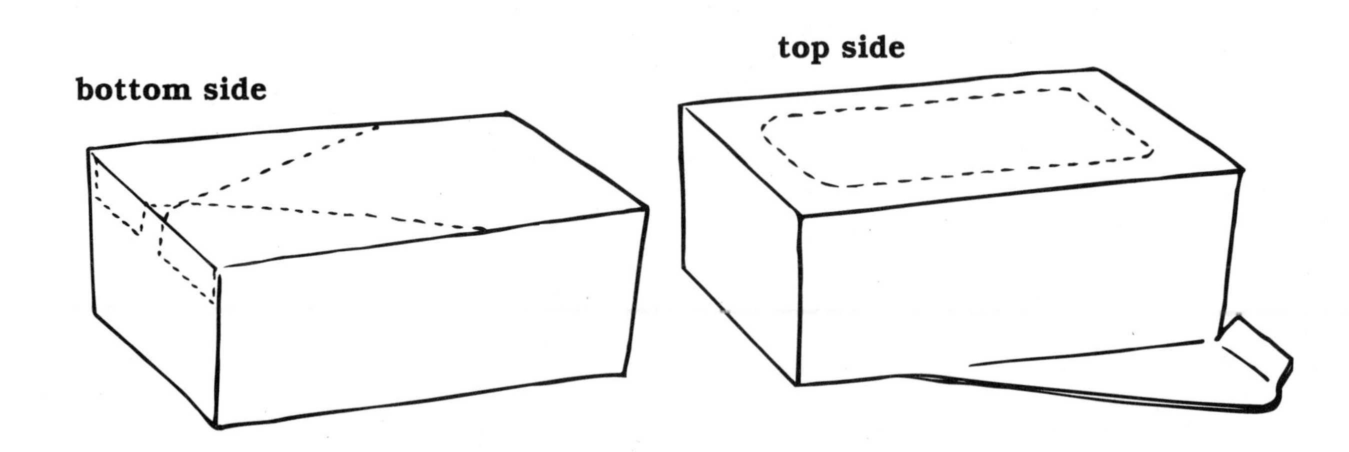

Space Shuttle

3. Once this side of the box is open, tape the loose end flaps to the outside of the box.
4. Turn the box over so the wings are on the floor.
5. Make an oval-shaped opening on the top side of the box.
6. Make a large ball with newspaper and masking tape, and tape the ball to the front end of the box for the nose cone.
7. Mix flour and water to make a papier-mâché paste.
8. Cut the newspaper into strips for children to dip into the paste and completely papier-mâché the space shuttle.
9. Once the papier-mâché has dried, poke two holes in four paper bowls and eight holes in the end of the box. Align the holes and attach the paper bowls with yarn to create burners on the back of the space shuttle.
10. Have children paint the shuttle white.
11. Once the paint has dried, children can paint the nose cone, burners, and other details using black tempera paint.
12. Children can glue red tissue "flames" to the burners.

Option:

- Use papier-mâché for the nose cone only. Paint the shuttle white.
- Color and cut out patterns (pp. 32, 33, 40) to glue to the shuttle.

Book Links:

- *The Earth and Sky - A First Discovery Book* created by Gallimard Jeunesse and Jean-Pierre Verdet (Cartwheel)
- *Floating Home* by David Getz (Holt)

Rocket Ship

Materials:

Large plastic trash can, large sheets of poster board, masking tape, yarn, tempera paint, sponges, shallow tins (for paint), star stickers (optional), sharp instrument for cutting (for adult use only)

Directions:

1. Turn the trash can upside down and cut a large opening in the side.
2. Cut three large triangles from poster board for wings.
3. Tape two pieces of poster board together, and cut one large circle.
4. Cut a slit to the center of the circle and form into a cone shape and tape. This becomes the nose cone.

Rocket Ship

5. Punch three sets of holes along the straight edge of each wing. Punch three sets of holes in the trash can for each wing and tie together according to the diagram below.

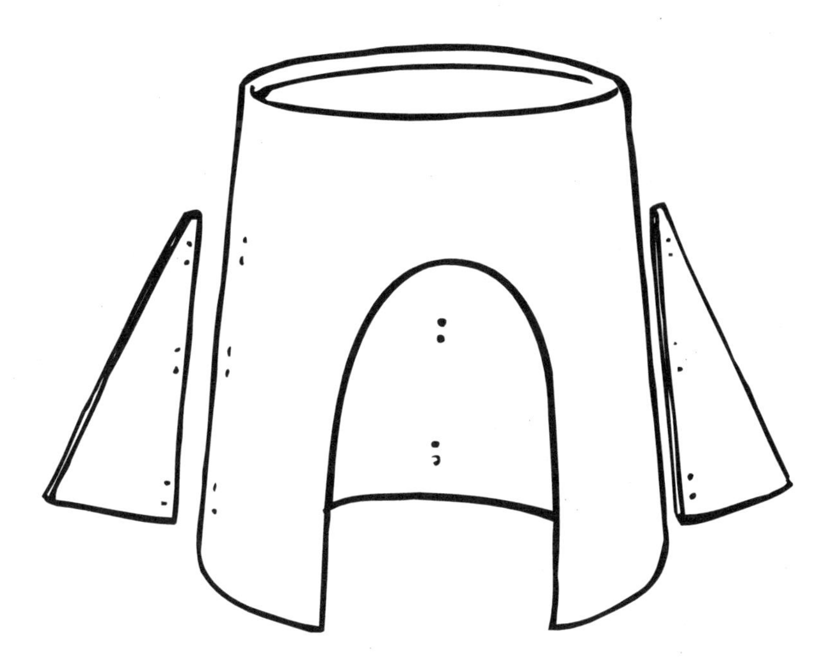

6. Punch holes in the nose cone and the trash can and secure the two with yarn.
7. Provide tempera paint and sponges for children to use to paint the wings and nose cone.
8. Let children decorate the rocket with star stickers.

Video Link:
- *There Goes a Spaceship* (Warner Vision)

Mars and the Rover

Materials:

Cereal boxes (one per child), large flat piece of cardboard, metal juice lids (six per child), foil, straws (six per child), tape, scissors, tempera paint (red and black), pencil, paintbrushes, combs, shallow tins (for paint), salt, flour, water, disposable bowls and spoons, small pebbles, red food coloring, ice pick or screwdriver (for adult use only)

Directions for Rover:

1. Punch one hole in each juice lid with ice pick or screwdriver.
2. Have children wrap their cereal boxes with foil.
3. Punch three holes on each side of each foil-wrapped box with a pencil or screwdriver. (Align holes.)
4. Have each child insert three straws completely through both sides of his or her box.
5. Cut slits on the ends of the straws.
6. Have children put a juice lid on each straw and tape the flaps of the straws to the juice lids.
7. Have children paint the top of their boxes black and use combs to make lines in the paint.

Directions for Mars Surface:

1. Mix salt, flour, water, and red food coloring in disposable bowls to make a thick, lumpy paste.
2. Have children coat the cardboard sheet with paste.
3. Let children add small pebbles to surface.
4. Once the surface is dry, children can use their rovers on Mars.

Book Link:

- *Mars* by Elaine Landau (Franklin Watts)

Planetarium

Materials:

Dryer box, black felt, black tempera paint, paintbrushes or rollers, shallow tins (for paint), glow-in-the-dark plastic self-adhesive stars and planets (available at nature stores, science stores, and novelty stores), gold and silver star stickers (optional), duct tape, flash-lights, sharp instrument for cutting (for adult use only)

Directions:

1. Completely remove one end of the box.
2. Cut an opening on one side of the box large enough for a child to crawl through.
3. Provide black tempera paint for the children to use to paint the inside and outside of the box.
4. Once the paint has dried, provide glow-in-the-dark, plastic, self-adhesive stars and planets for the children to use to decorate the inside of the box.
5. Children can use gold and silver star stickers to decorate the outside of the box.
6. Tape a piece of black felt to the inside of the box to cover the opening. This will make it dark inside the box.
7. Place the box upside down in the sunlight for an hour to activate the glow-in-the-dark stars and planets.
8. Children can use flashlights inside the planetarium

Book Links:

- *The Glow-in-the-Dark Zodiac* by Katherine Ross (Random House)
- *How Many Stars in the Sky?* by Lenny Hart (Mulberry)

Life-Size Stuffed Astronaut

Materials:

Space Helmet (p. 11), Oxygen Pack (p. 12), three-foot (.9 m) dowel, white shirt and pants, pillow case, boots, gloves, newspaper, yarn, safety pins, markers

Directions:

1. Have children stuff the pillow case, shirts, pants, and gloves with newspaper.
2. Tie the pillow case to the top of the dowel with yarn.
3. Insert the dowel through the shirt.
4. Place the pants on the bottom of the dowel, and tie together with yarn.
5. Stuff the bottom of the pant legs into the boots.
6. Pin the gloves onto the end of the shirt sleeves.
7. Provide markers for the children to use to add facial features and hair to the astronaut.
8. Place the oxygen pack on the astronaut's back and the space helmet on the astronaut's head.

Book Links:

- *I Am an Astronaut* by Cynthia Benjamin (Barron's)
- *The Astronauts* by Dinah L. Moche (Random House)

Space Helmet

Materials:

Empty cereal box (double-package size) or paper grocery bag, pipe cleaners, tissue paper, liquid starch, paintbrushes, shallow tins (for liquid starch), hole punch, pipe cleaners, space stickers (optional), sharp instrument for cutting (for adult use only)

Directions:

1. Remove the open flaps on the top of the box.
2. Cut a large semi-circle in the front side of the box.
3. Punch a hole on each side of the opening.
4. Cut tissue paper into various shapes and sizes.
5. Children brush liquid starch on the helmet and stick on tissue paper pieces.
6. Once the tissue paper has dried, children can decorate the helmet with stickers and attach pipe cleaners through the holes.

Note:

These directions are for making one helmet. You can make several for children to share, or have each child bring in a cereal box from home to make his or her own helmet.

Book Link:

• *I Want to Be An Astronaut* by Byron Barton (Harper Trophy)

Oxygen Pack

Materials:
Rectangular facial tissue box, white butcher paper, tape, elastic, black tempera paint, gadgets for stamp painting (potato masher, wire egg whip, spools, blocks, cookie cutters), shallow tins (for paint), scissors

Directions:
1. Wrap the facial tissue box with butcher paper.
2. Cut an opening in the butcher paper corresponding with the opening in the tissue box.
3. Punch two holes on opposite sides of the box.
4. Provide tempera paint and gadgets for the children to use to stamp paint the oxygen pack.
5. Once the paint has dried, insert the elastic through the holes and tie the loose ends together in a knot. A child's arms go through the elastic straps and the pack is worn on the back.

Note:
These directions are for making one oxygen pack. You can make several for children to share, or have each child bring in a tissue box from home to make his or her own pack.

Book Links:
- *One Giant Leap* by Mary Ann Fraser (Holt)
- *1000 Facts About Space* by Pam Beasant (Kingfisher)

Paper Space Shuttle

Materials:

Space Shuttle Patterns (pp. 14-15), sturdy paper, old file folders, scissors, tape, crayons or markers

Directions:

1. Duplicate the space shuttle patterns, cut out, and attach the halves with clear tape.
2. Trace the space shuttle patterns onto sturdy paper and cut out. Make several for children to use as templates.
3. Have children trace the space shuttle patterns onto old file folders and cut out.
4. Cut slits in the body and the wing of each space shuttle according to the diagrams below.

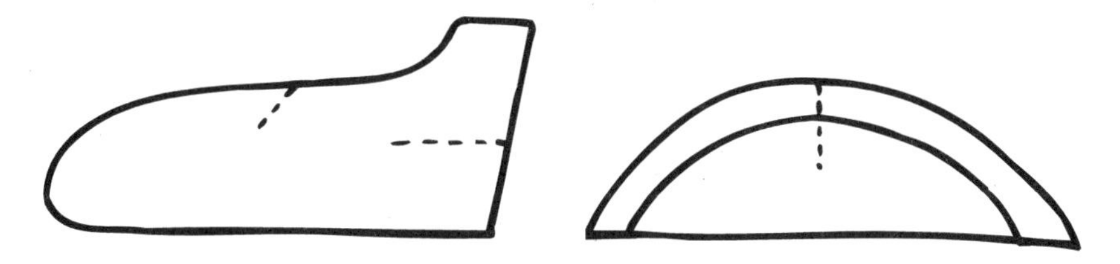

5. Children can use crayons or markers to color the shuttles.
6. Demonstrate how to insert the wing through the slit and tape the end.

Book Link:

• *Let's Go to the Moon* by Janis Knudsen Wheat (National Geographic Society)

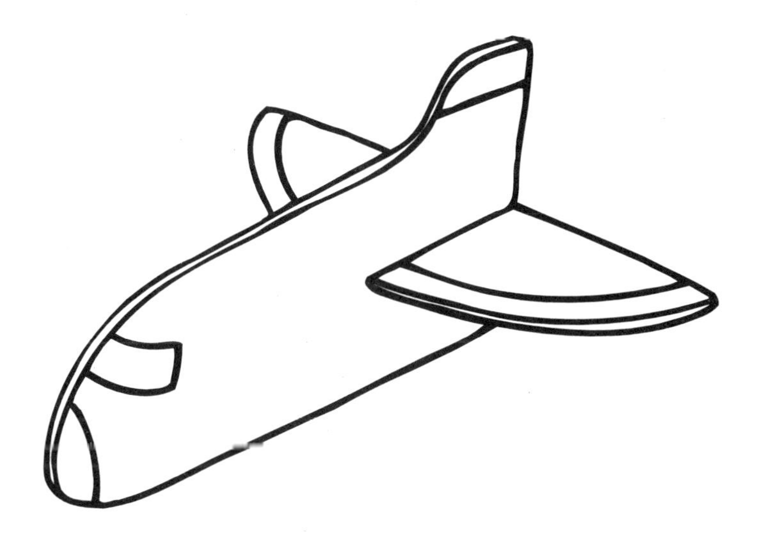

Space Shuttle Patterns

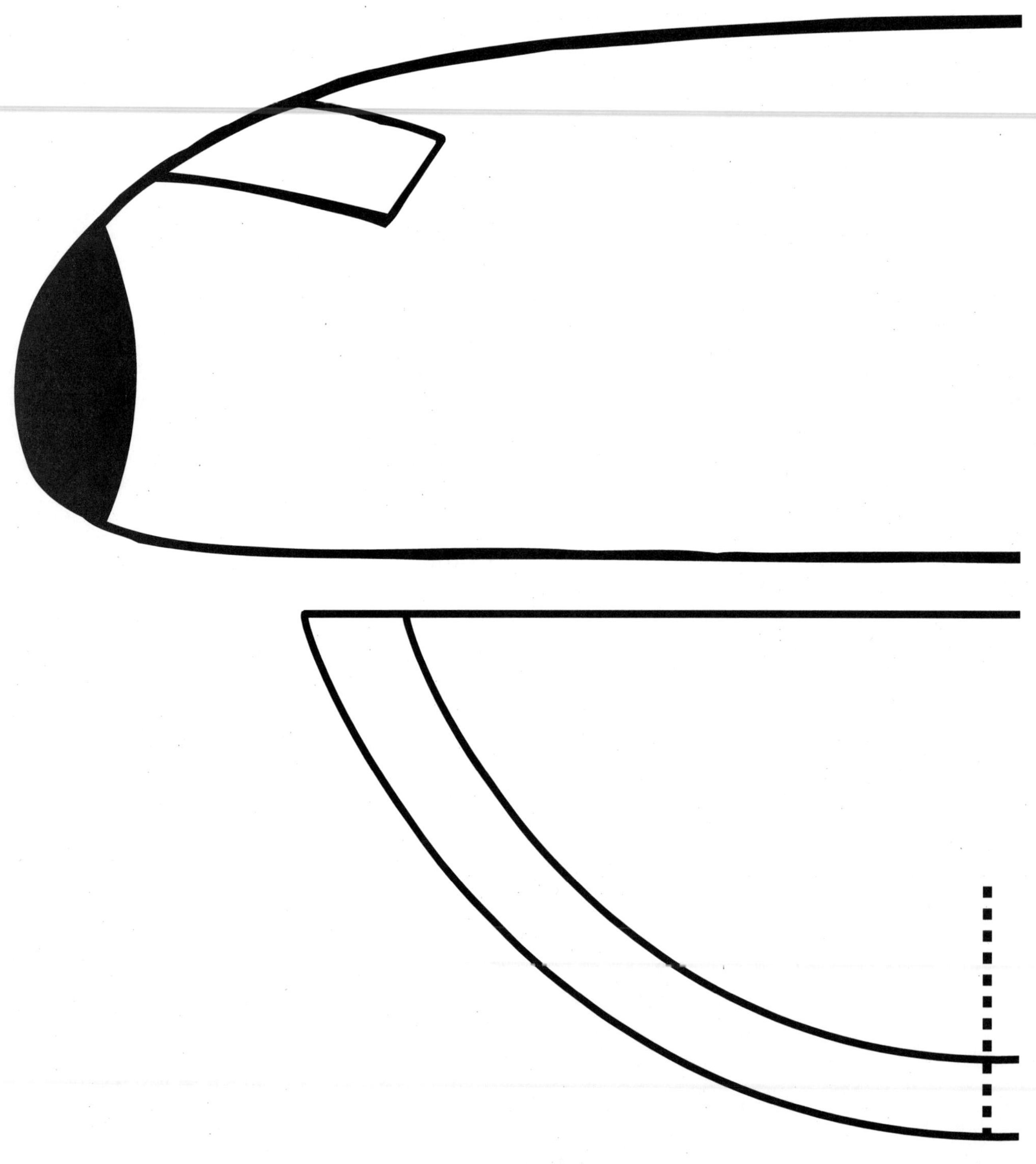

Space Shuttle Patterns

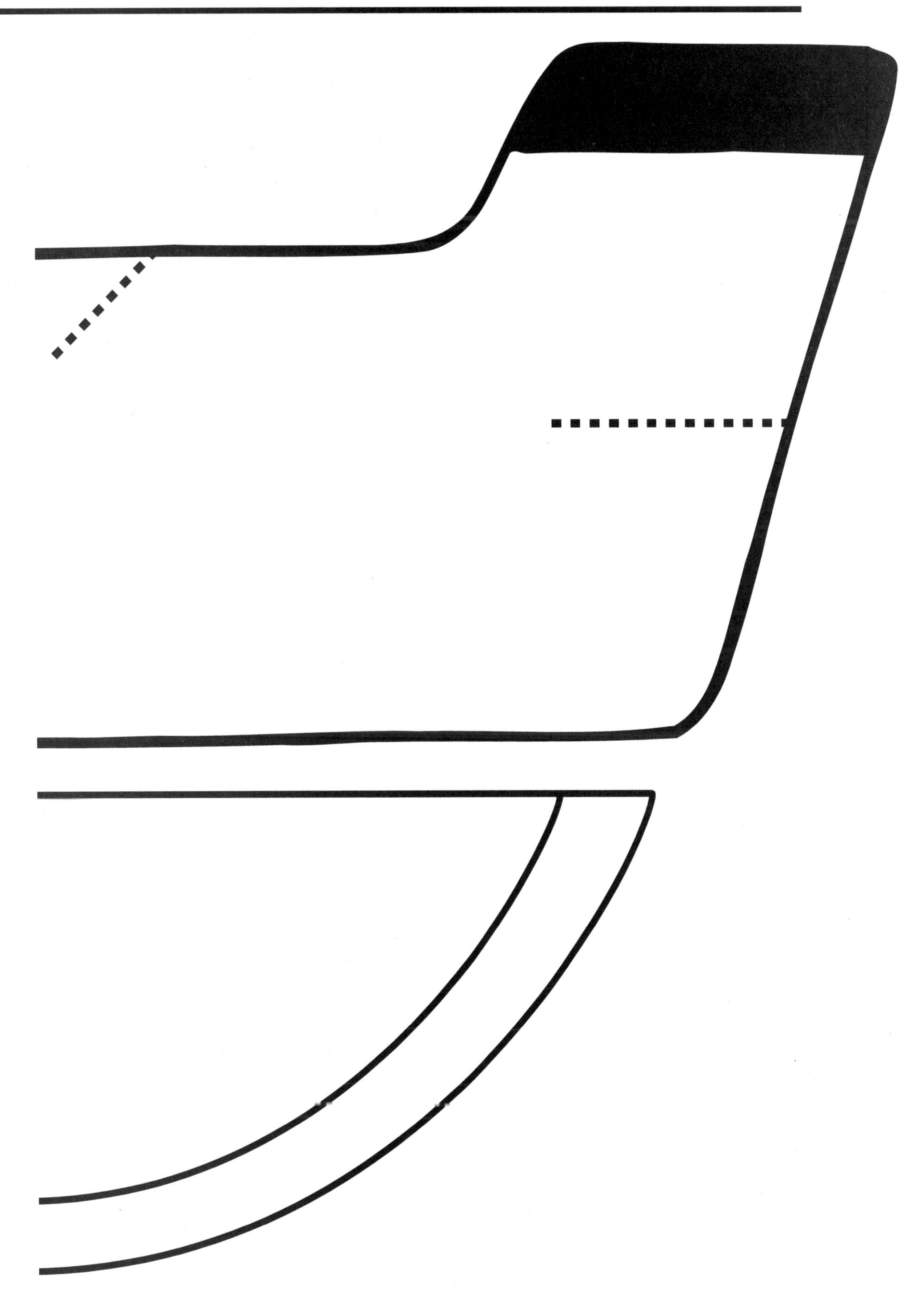

Astronaut Paper Doll

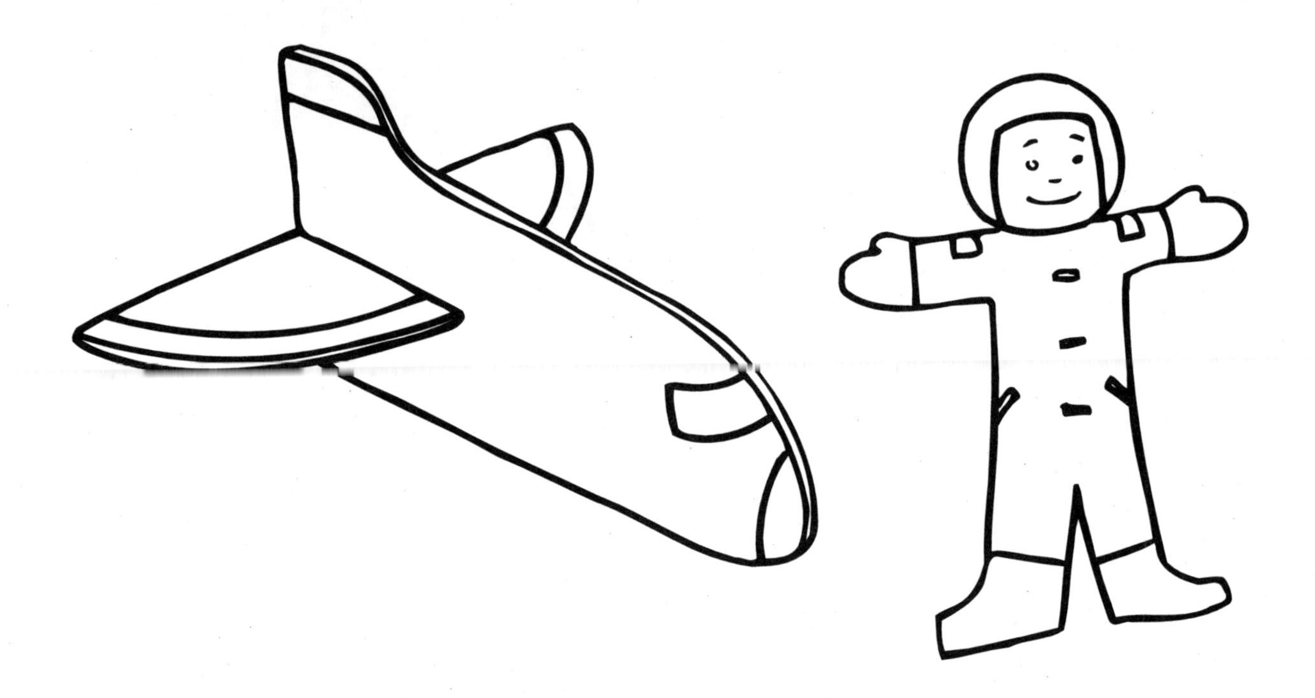

Materials:
Paper Doll Pattern (p. 17), Paper Doll Clothing (p. 18), sturdy paper, old file folders, white construction paper, paper clips, scissors, tape, crayons or markers

Directions:
1. Trace the doll pattern onto an old file folder and cut out. Make one for each child.
2. Trace the clothing patterns onto construction paper. Make one set for each child.
3. Provide crayons and markers for children to use to draw facial features and other desired details on their dolls.
4. Children cut out the clothing and attach it to their dolls with paper clips.
5. Children can insert their paper dolls into the top slits of the space shuttles.

Option:
- Children can store their dolls and clothing patterns in envelopes.

Video Link:
- *Moon Man* (Children's Circle)

Paper Doll Pattern

Paper Doll Clothing

Rocket Tube

Materials:

Rocket Tube Patterns (p. 20), paper towel tube (one per child), construction paper (in assorted colors), sturdy paper, tempera paint, paintbrushes, shallow tins (for paint), scissors, stapler, glue, star stickers (optional)

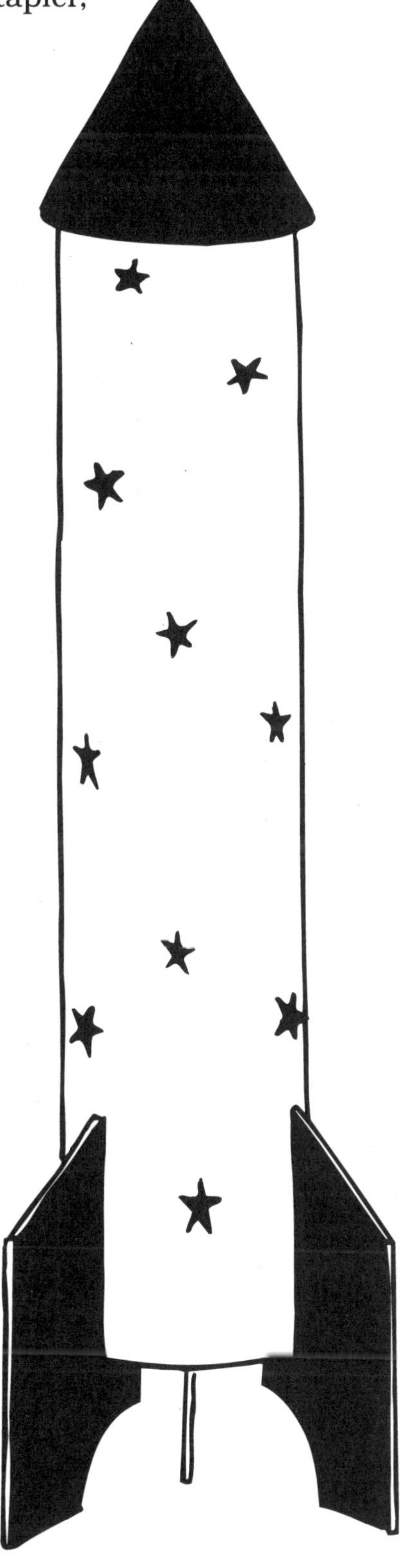

Directions:

1. Trace the rocket tube patterns onto sturdy paper. Make several for children to use as templates.
2. Provide tempera paint for the children to use to paint their paper towel tubes.
3. Have children trace the nose cone onto construction paper, cut out, fold, and staple.
4. Have each child trace the wing pattern three times onto sturdy paper and cut out.
5. Once the tubes have dried, children can decorate them with star stickers.
6. Help children cut three slits in one end of their tubes and insert the wings.
7. Glue a nose cone on the top of each rocket tube.

Book Link:

• *Magic School Bus: Lost in the Solar System* by Joanna Cole (Scholastic)

Rocket Tube Patterns

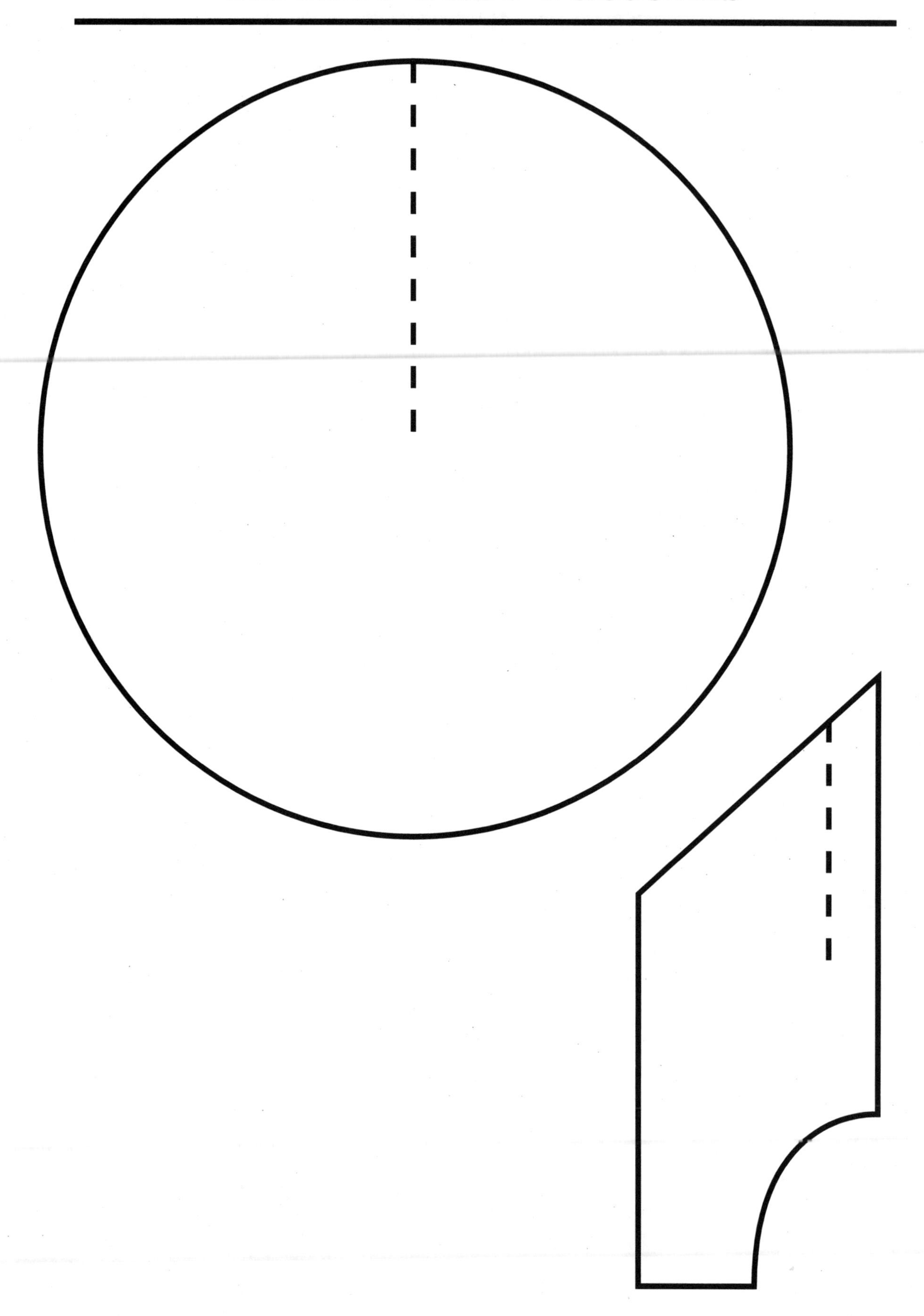

Moon Craters

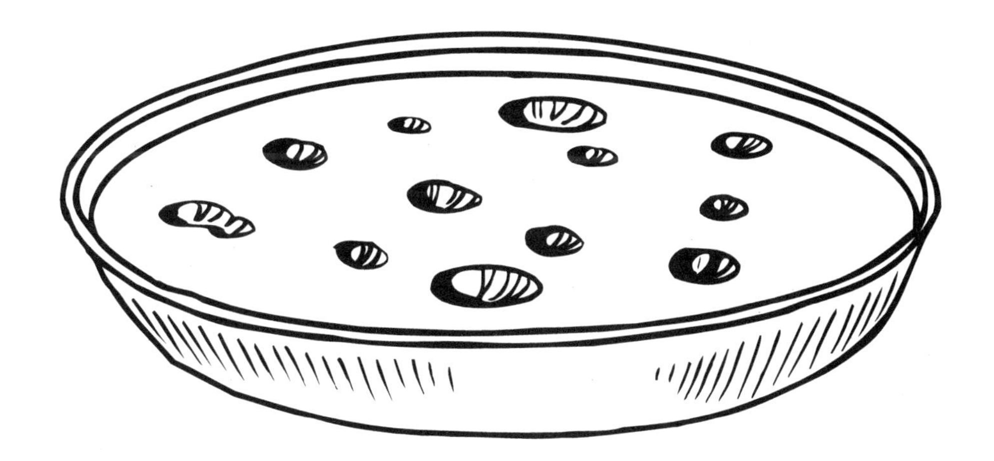

Materials:

Plaster of Paris, aluminum pie tin or Styrofoam bowl (one per child), pebbles or marbles

Directions:

1. Mix the plaster of Paris according to directions on the package, and pour some into each aluminum pie pan or Styrofoam bowl.
2. Provide pebbles or marbles for the children to place in the plaster of Paris to form moon craters.
3. Once the plaster of Paris has dried, children can remove the pebbles or marbles.
4. Remove the moons from the pans or bowls.

Book Links:

- *Happy Birthday Moon* by Frank Asch (Prentice-Hall)
- *Monkey and the Moon* by John Randall (Abbeville)
- *Moongame* by Frank Asch (Prentice-Hall)

Universe Necklace

Materials:
Bread dough ingredients (see recipe below), clear straws, tempera paint (in assorted colors), shallow tins (for paint), cotton swabs, yarn, yarn needles, scissors

Directions:
1. Cut a long piece of yarn for each child.
2. Cut some of the straws into short sections.
3. Make bread dough with the children.
4. Show children how to make small balls out of the bread dough and insert a long straw through the balls. Let the dough dry.

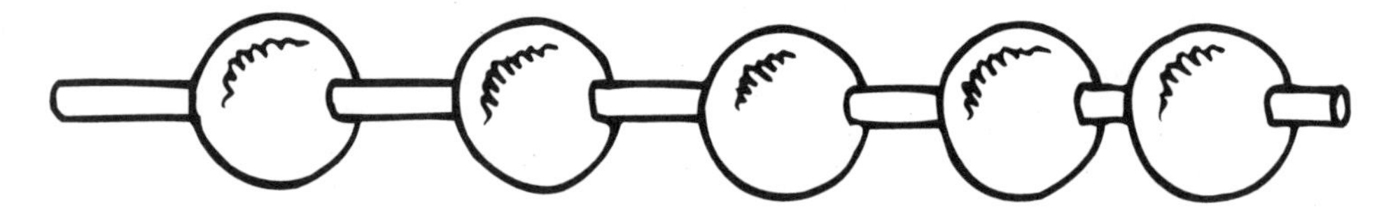

5. While the dough is still on the straw, have children paint the balls with cotton swabs and tempera. They can paint the beads to look like the sun, moon, and planets.
6. Once the paint has dried, carefully remove the beads from the straws.
7. For each child, tie a section of straw to one end of a yarn length and the needle to the other.
8. Have the children thread the planets and the straw sections to make necklaces.
9. Once a necklace is completed, remove the needle and tie the loose pieces of yarn together.

Recipe:
1 cup (.25 kg) salt
1/2 cup (.13 kg) cornstarch
1/2 cup (.13 l) water (heated)

Dissolve the salt in the heated water. Stir in cornstarch. Knead until smooth

Book Link:
- *Is There Life in Outer Space?* by Franklin M. Branley (Harper Trophy)

Papier-Mâché Planets

Materials:

Newspaper, masking tape, papier-mâché paste (flour and water), tempera paint, sponge pieces, shallow tins (for paint), scissors

Directions:

1. Cut newspaper into strips.
2. Have children work together to make a large ball from newspaper and masking tape.
3. Mix flour and water to make a papier-mâché paste.
4. Provide newspaper strips for the children to dip into the paste and use to cover the large newspaper ball.
5. Once the papier-mâché has dried, children can paint the planet.

Options:

- Make several planets to hang around the classroom.
- Make Saturn's rings by cutting the center out of a paper plate and placing it over the planet. (The circle must be snug enough to stay on.)

Book Link:

- *Looking at the Planets: A Book About the Solar System* by Melvin Berger (Scholastic)

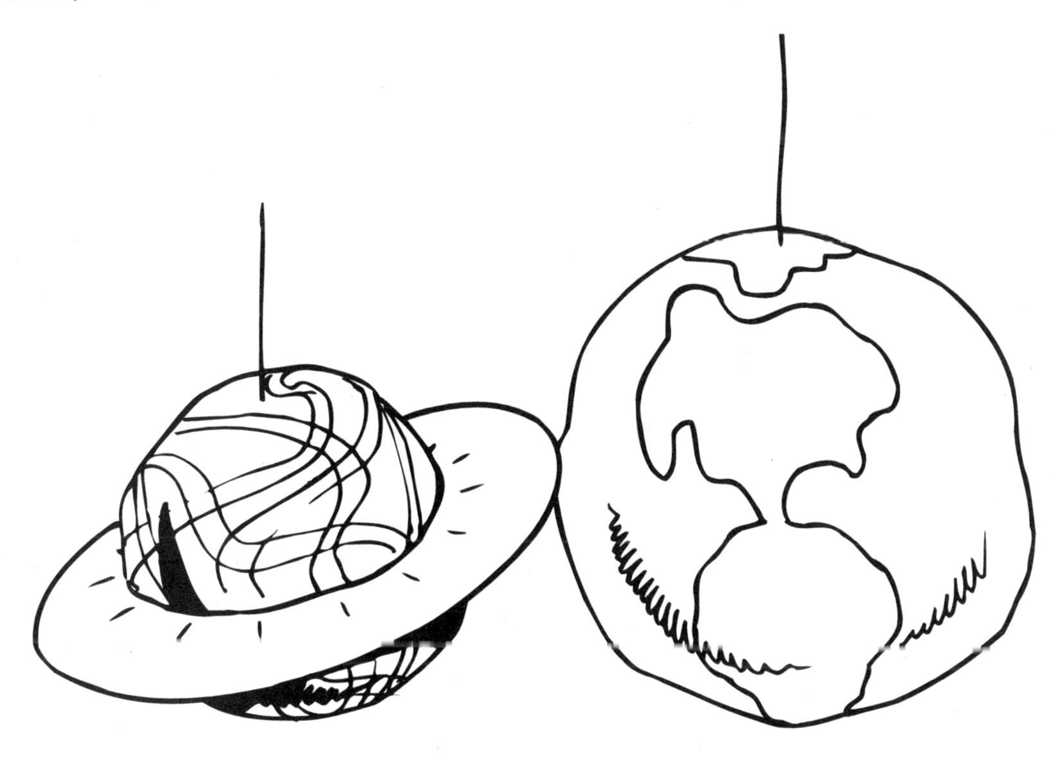

Three-Dimensional Earth

Materials:

Papier-mâché planet (p. 23), plastic bubble packing wrap, globe, construction paper (brown and green), blue tempera paint, shallow tins (for paint), glue, markers, scissors, tape

Directions:

1. Cut the plastic bubble wrap into small sections.
2. Looking on a globe, outline the continents on construction paper and cut out.
3. Children can dip the bubble wrap in the blue tempera and stamp paint the oceans directly onto the papier-mâché planet.
4. Children can rip the construction paper into small sections and glue it onto the continents to give them texture.
5. Glue the continents to the Earth.

Book Link:

- *The Third Planet: Exploring the Earth from Space* by Sally Ride and Tam O'Shaughnessy (Crown)

Paper Plate Sun

Materials:

Large lightweight paper plates (one per child), Popsicle sticks, yellow tempera paint, shallow tins (for paint), plastic bubble packing wrap, scissors

Directions:

1. Cut pairs of slits around the paper plate, as shown.
2. Have children insert Popsicle sticks through the slits.
3. Provide small sections of bubble wrap for children to use to stamp paint the sun.

Options:

- Children can paint both sides of their suns.
- Hang the suns from the ceiling using clear wire.
- Provide gold foil for children to use to decorate their suns.

Book Link:

- *What Is the Sun?* by Reeve Lindbergh (Candlewick)

Sewing Saturn

Materials:

Lightweight paper plates (two per child), construction paper (in assorted colors), yarn, yarn needles, scissors, hole punch, water-colors

Directions for Each Saturn:

1. Fold a piece of construction paper in half, trace two circles on the fold (according to the diagram below), and cut out. (Make sure the ring is large enough to fit around the paper plates.)

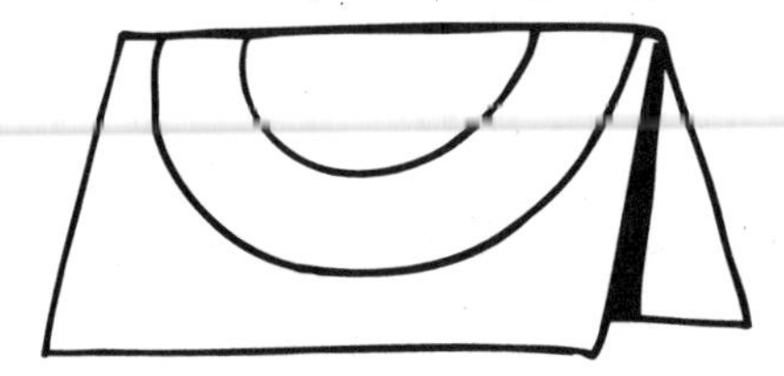

2. Place two paper plates together and punch holes around the edges. (Make sure the holes are aligned.)
3. Turn the plates so they are inside-to-inside.
4. Tie the yarn needle to a long piece of yarn, and tie the other end through two aligned holes on the paper plates.
5. Each child stitches around his or her joined paper plates.
6. Once the stitching is completed, remove the yarn needle and tie the loose piece of yarn to the plates.
7. Each child can paint his or her Saturn.
8. Once the watercolors have dried, help each child insert the Saturn through the ring.

Option:

• Suspend the Saturns from the ceiling.

Book Links:

• *Saturn* by Don Davis and Ian Halliday (BLA)
• *Saturn: The Spectacular Planet* by Franklyn M. Branley (Crowell)

Jupiter Jars

Materials:
Small jars with lids (one per child), salt or sand, disposable bowls, powdered tempera paint (in assorted colors), transparent tape, spoons

Directions:
1. In disposable bowls, mix small amounts of powdered tempera paint with salt or sand. Make batches of different colors.
2. Children spoon the salt or sand mixtures into the jars, alternating layers of colors to simulate Jupiter's belts.
3. Place a lid on each jar and secure by wrapping with transparent tape.

Book Links:
- *Jupiter* by Seymour Simon (William Morrow)
- *Jupiter: The Spotted Giant* by Isaac Asimov (Gareth Stevens)

Planet Pluto

Materials:
Round balloons, yarn, liquid starch, shallow tins (for starch), scissors, straight pin (for adult use only)

Directions:
1. Cut yarn into long sections.
2. Blow up one balloon for each child.
3. Have children dip the yarn into the starch, squeeze the excess with their fingers, and place the yarn strands on the balloons.
4. Once the starch has dried, pop each balloon. Make sure that all balloon pieces are thrown away!
5. Suspend the planet Plutos from the ceiling.

Book Links:
- *Mysteries of the Planets* by Franklyn M. Branley (Dutton)
- *Pigs in Space - Journey to the Planet ZA* by Kate Foster (Grosset and Dunlop)

Star Dancer

Materials:

Star Patterns (p. 30), construction paper, straws, yarn, yarn needles, marker, scissors, hole punch

Directions:

1. Trace the star patterns onto construction paper. Make one set per child.
2. Cut one long piece of yarn for each child.
3. Cut the straws into short sections.
4. Tie a section of straw to one end of each yarn length and a yarn needle to the other.
5. Have children cut out the stars, and punch holes in the center of each star.
6. Have children thread the straws and stars (sequencing the stars from large to small).
7. Children end their star dancers with a straw.
8. Remove the needles for the children, and tie the loose ends of yarn to the straws.
9. Star Dancers can be hung as mobiles.

Book Links:

- *I Wonder Why Stars Twinkle and Other Questions About Space* by Carole Scott (Kingfisher Books)
- *Stargazers* by Gail Gibbons (Holiday House)
- *Stars* by Michael George (Creative Education)

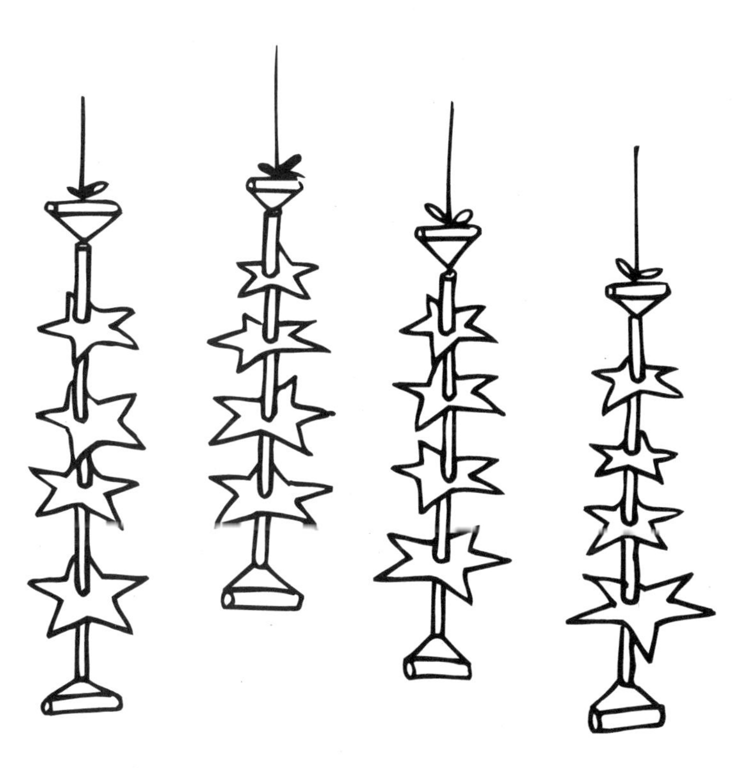

Star Dancer Patterns

Space Mobile

Materials:

Space Mobile Patterns (pp. 32-33), old file folders or sturdy paper, white construction paper, crayons or markers, yarn, scissors, hole punch, stapler, star stickers and self-sticking dots (optional)

Directions:

1. Cut old file folders length-wise into four strips. Cut enough so that each child has one strip.
2. Trace the space patterns onto sturdy paper and cut out. Make several for the children to use as templates.
3. Have children trace the space patterns onto white construction paper and cut out.
4. Children decorate the space objects and file folder strips with crayons, markers, and stickers.
5. Have children punch seven holes in their file folder strips (making sure there is space between each hole) and one hole in each of the space objects.
6. Attach the space objects to the file folder strips by tying with yarn.
7. Fold each folder strip into a circle and staple.
8. Space mobile can be displayed by punching two holes at the top of each and hanging with a piece of yarn.

Book Links:

- *Night Sky* by Carole Stott (Dorling Kindersley)
- *Stars in the Sky* by Allan Fowler (Children's Press)

Space Mobile Patterns

Space Mobile Patterns

Star Catcher

Materials:

Butcher paper, black tempera paint, squeeze bottles, shower squeegees, salt, salt shakers, glue, scissors, hole punch, tape or yarn, newsprint

Directions:

1. Spread newsprint over the work station.
2. Cut butcher paper into a long section.
3. Have children squeeze black tempera paint onto the butcher paper and then use the shower squeegees to paint the paper black.
4. Once the paint has dried, cut the butcher paper into smaller sections. Make one section per child.
5. Have children fold their papers into two to four sections and punch holes in the papers.
6. Have the children unfold their papers, drizzle the papers with glue, and then sprinkle the papers with salt.
7. Once the glue has dried, tape the Star Catchers to a window or use yarn to hang them in front of a window.

Options:

- If shower squeegees are not available, children can use roller brushes to paint the paper black.
- Sheets of thin black paper can be substituted for butcher paper painted black.

Book Link:

- *The Sky is Full of Stars* by Franklyn M. Branley (Harper Trophy)

Asteroid Painting

Materials:

Butcher paper, socks, knee-high nylons, sand, rice, beans, oatmeal, rubber bands, measuring cup, tempera paint (in assorted colors), shallow tins (for paint), masking tape, smocks (optional)

Directions:

1. Fill socks and knee-high nylons with one scoop of sand, rice, beans, or oatmeal.
2. Secure the top of each sock or nylon with a rubber band. Be sure to leave space between the ingredient and the rubber band.
3. Cut a long strip of butcher paper and secure it to the floor with masking tape.
4. To make asteroid prints, children dip the sock or nylon in tempera paint and drop the sock or nylon onto the butcher paper.
5. Post the finished mural on a wall.

Puzzle Link:

- *Space Adventure* (Small World Toys)

Book Link:

- *Comets, Meteors, and Asteroids* by Seymour Simon (Morrow Junior Books)

Cloud Painting

Materials:
Sturdy paper, detergent flakes (like Ivory Snow), water, large bowl, manual egg beater, spoon, scissors, hole punch, yarn, newspaper

Directions:
1. Cut large cloud shapes from sturdy paper. Make one for each child.
2. Pour detergent flakes and water into a large bowl.
3. Let children take turns using an egg beater to make a thick mixture for finger painting.
4. Spread newspaper over the work stations.
5. Have children spoon the detergent onto their clouds and then finger paint designs on the clouds. (Remind children not to touch eyes with soapy hands.)
6. Once the clouds have dried, punch a hole in each one and string with yarn to hang.

Book Links:
- *It Looked Like Spilt Milk* by Charles G. Shaw (HarperCollins)
- *Little Cloud* by Eric Carle (Philomel)

Playdough Planets

Materials:
Playdough ingredients (see recipe below),
rolling pin, cooking utensils, plastic knives, paper plates

Directions:
1. Make the playdough with the children, using caution when adding the boiling water.
2. Once the playdough has cooled, have the children create planets, stars, sun, moon, asteroids, and so on from the dough.
3. Children can keep their miniature universes on paper plates.

Playdough Recipe
4 cups (1 kg) flour
2 cups (.5 kg) salt
8 tsp. (40 g) cream of tartar
10 tsp. (50 ml) liquid vegetable oil
4 cups (1 l) boiling water
food coloring (desired colors)

Directions:
1. Combine the first four ingredients in a large bowl.
2. Add food coloring to the boiling water.
3. Pour the water into the dry ingredients and mix.
4. Remove the dough from the bowl and knead on a floured surface.

Crayon Resist Skies

Materials:
White construction paper, sturdy paper, crayons, black tempera paint, paintbrushes, shallow tins (for paint)

Directions:
1. Make a watery mixture of black tempera paint.
2. Cut several different-sized circles from the sturdy paper. These will become planet templates for children to trace.
3. Have children trace the circles onto white construction paper.
4. Children can use crayons to color the planets. Have them press very firmly when coloring.
5. Provide the watery black tempera for children to use to wash over their entire pictures.

Options:
- Children can draw stars on their pictures before painting with tempera.
- Provide star stickers for children to add after the paint has dried.

Puzzle Link:
- *Our Solar Syste*m - floor puzzle (Frank Schaffer)

Solar System Matching

Materials:

Solar System Activity Sheet (p. 40), construction paper (in assorted colors), crayons or markers, glue stick, scissors

Directions:

1. Duplicate a copy of the solar system activity sheet and cut out the patterns.
2. Trace the patterns onto assorted colors of construction paper and cut out. (Use a variety of colors for each size and shape.)
3. Duplicate a copy of the solar system activity sheet for each child.
4. Have children match the patterns by size and shape and glue them onto the solar system activity sheets.
5. Children can use crayons or markers to decorate the sheets.

Option:

• Cut out two sets of the solar system shapes for each child. Have children match the shapes together.

Solar System Activity Sheet

Astronaut Letter Match

Materials:
Game Board Pattern (pp. 42-43), Astronaut Game Cards (p. 44), colored markers or pencils, clear contact paper, scissors, clear tape

Directions:
1. Duplicate the game board pattern and color, as desired.
2. Attach the two pieces of the game board using tape.
3. Cover the game board pattern with clear contact paper, or laminate.
4. Duplicate the game cards, color, cut apart, laminate, and cut apart again. Be sure to leave a thin laminate border to prevent peeling.
5. Children spell the word astronaut by matching the game cards to the astronauts on the game board.

Option:
• Make one set of cards and one copy of the game board for each child. Let the children color their game boards and cards.

Game Board Pattern

Game Board Pattern

Astronaut Game Cards

Rocket Ship Countdown

Materials:
Rocket Pattern (p. 46), crayons, markers or colored pencils, glue stick, scissors

Directions:
1. Duplicate two copies of the rocket pattern for each child.
2. For each child, cut the numbers of one rocket pattern into sections.
3. Have children match and glue the numbers to the rocket ships.
4. Provide crayons, markers, or colored pencils for children to use to add details to their pictures.

Option:
• White-out the numbers from one copy, duplicate, and have children glue the number sections on in sequence.

Rocket Pattern

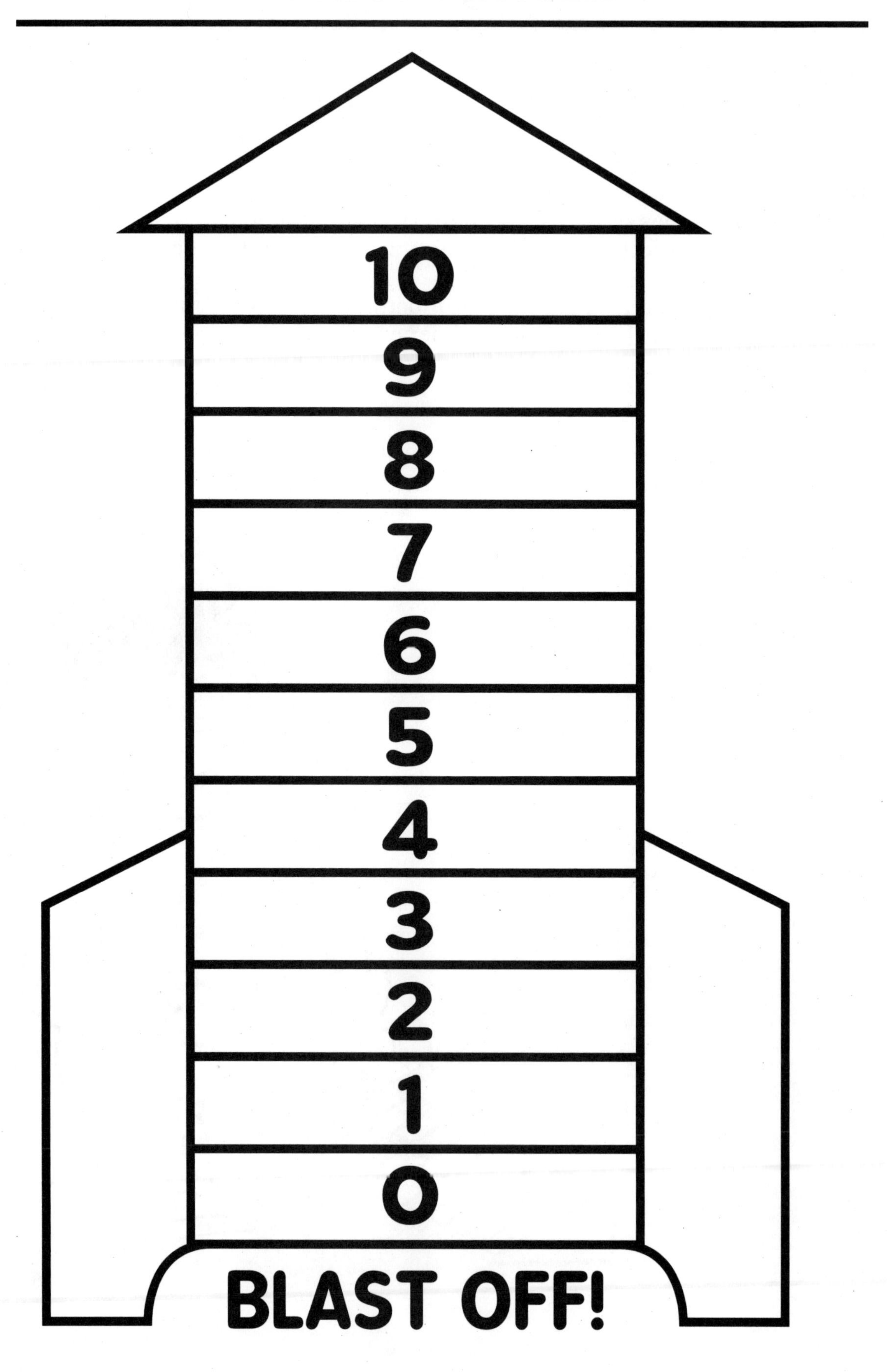

Space Memory Match

Materials:
Space Patterns (p. 48), colored markers, scissors, clear contact paper

Directions:
1. Duplicate the space patterns twice, color, cut apart, cover with contact paper or laminate, and cut out again. (Leave a thin laminate border around each pattern to help prevent peeling.)
2. Shuffle the cards and spread them face down on a table.
3. Demonstrate how to play the game. The object is to match the space patterns by turning the cards over two at a time. If a match is made, the cards remain face up and the child takes another turn. If a match isn't made, the cards are turned over and the next child takes a turn. Game continues until all cards are face up.

Option:
• Introduce the game by leaving the shuffled cards face up and having the children simply match the space patterns together.

Space Patterns